This how-to book was written as an instructional guide to diagnose, unclog, and prevent common household drain line stoppages, based on the author's knowledge and trade skills of the sewer and drain cleaning business. The publisher and the author disclaim any personal liability, loss, or risk incurred as a result of any information or advice contained herein, either directly or indirectly. Furthermore, the publisher and author do not guarantee that the holder of this information will make profit from the information contained herein. Any mention of promises to make money, either implied or not implied, is strictly based on the author's opinion of the information contained herein.

The opinions expressed in this manuscript are solely the opinions of the author and do not represent the opinions or thoughts of the publisher. The author has represented and warranted full ownership and/or legal right to publish all the materials in this book.

Drain Cleaning 101
All Rights Reserved.
Copyright © 2013 Tony Marini
v6.0

Cover design by Michael Hamboussi. All rights reserved - used with permission.

This book may not be reproduced, transmitted, or stored in whole or in part by any means, including graphic, electronic, or mechanical without the express written consent of the publisher except in the case of brief quotations embodied in critical articles and reviews.

Outskirts Press, Inc.
http://www.outskirtspress.com

ISBN: 978-1-4787-1935-9

Outskirts Press and the "OP" logo are trademarks belonging to Outskirts Press, Inc.

PRINTED IN THE UNITED STATES OF AMERICA

Greetings,

Let me take a moment to introduce myself. My name is Tony Marini and I have over 23 years of experience in the sewer and drain cleaning industry. My company has been providing sewer and drain cleaning services within the five boroughs of New York to property management companies with multi-family dwellings, as well as to commercial establishments and residential homeowners since 1996. I am a professional in the trade, and my reputation in the business speaks for itself.

I have taken the liberty of writing this step-by-step instructional book to explain the proper techniques used to diagnose, unclog, and prevent common drain line stoppages. This book will benefit the beginner, intermediate, and even the expert drain cleaner.

Drain cleaning is not difficult. If you have the proper knowledge, tools, confidence in yourself, and patience, you can unclog common drain line stoppages like a pro, by doing it yourself.

You will not find a more comprehensive do-it-yourself book than this one. In my research I have come across very basic techniques on how to unclog drain line stoppages, and the most common technique I have found was by using a plunger. Why would you want to do this? Yes, a plunger may temporarily unclog a drain line stoppage by pushing it further down the line, but this can cause an additional blockage later on.

The only plumbing fixture upon which a plunger should be used is the toilet. It's okay to knock the blockage down the toilet line, as this line is large enough to handle the stoppage. However in drain lines such as sinks, bathtubs, and other small drains,

I wouldn't recommend using a plunger, as the forced air can separate drain line connections, causing leaks.

When unclogging common drain line stoppages, the idea is to remove the stoppage from the drain line, not to push it further down. I will explain the techniques and tools that I use on a daily basis in unclogging all basic household drain line stoppages, while breaking down each step into an easy-to-follow process that you can duplicate.

Who can benefit from this material? .... Almost anyone!

- Perhaps you are looking to start your own business.
- Maybe you're a property management company with multi-family dwellings who is interested in minimizing cost and maximizing productivity of your superintendents and handymen, by having them unclog any drainage problems that may occur in your rental units.
- Are you a homeowner and wish to do it yourself? This book will explain how.
- Possibly you're a drain cleaning professional looking for some helpful tips or a reference guide to supply to new employees. This book will surely bail them out if they get stuck on a job.

I am certain that by using my step-by-step methods, you will be able to either SAVE MONEY or MAKE MONEY by learning how to diagnose and unclog common drain line blockages like a professional.

Sincerely yours,

Tony Marini - *Drainage Specialist*

Contents

Self-Preparation: 1
Work Area Safety: 2
Types of Drain Cleaning Machines 3
Drum Machine vs. Hand Drill Snake 6
Basic Use of Your Drain Cleaning Machine..............................9
Drain Cleaning Technique Comparison: 10
Drainage System Terminology: 12
Types of Drain Line Blockages: How to diagnose,
    unclog, and prevent future blockages..............................16
Combination Stack Lines, aka Combo Stack Line:.................73
Drain Cleaning Cable Sizes and Where They Are Used.........78
Recommended Hand Tools:...79
24-Hour Live Telephone Support 82

Self-Preparation:

BEFORE STARTING, MAKE sure that you are not wearing any loose or baggy clothing or jewelry; make sure your clothing fits as close to the body as possible but comfortably, since you are working with professional drain cleaning equipment. Its components spin quite fast when in operation mode, which may snag any loose-fitting clothing or jewelry. You want to be safe.

You do not want to get yourself caught up in the motion of the spinning drum or cable. Always remember to wear a pair of "ugly gloves" while operating the machine and when feeding its cable into a drain line, as the "ugly gloves" are specially designed for the use of handling drain cleaning cables. Don't forget to always wear safety glasses or goggles.

Work Area Safety:

JUST AS IT'S important to keep yourself neat and organized, your work area should be neat and organized as well. Yes, drain cleaning can be a bit of a messy job, especially if you have an overflowing plumbing fixture, but since we're discussing common household blockages, it's very unlikely, as the situation can be controlled. Always remember to keep your work space clutter free and allow yourself enough room to maneuver. When working it's best to keep a clean, comfortable, quiet environment so that you can concentrate on what you are doing.

For example: When working on a bathroom sink stoppage, remove any mats or objects from the floor that may be in your way; also remove all items from under and on top of the sink and place them outside of your work area. Place a work mat or small drop cloth on the floor by the sink, and set up your drain cleaning machine on the mat, which will help prevent any mess or property damage. If possible, close the door so you are alone and can concentrate on your project. It is also a good idea to keep small children and pets out of your work area.

Types of Drain Cleaning Machines

THE TYPES OF drain cleaning machines that I use in my business on a daily basis are manufactured by the Ridge Tool Company. After using many different types of drain cleaning machines, in my opinion, the Ridgid® brand prevailed over all the rest. They have proven to be the safest and easiest to maneuver and operate, especially for a single operator, and they get the job done right.

The drain cleaning machines that I currently use are:

Ridgid® K-3800 Drum Machine – which is excellent for cleaning ¾" - 4" lines, and can operate four different-sized cables; ¼", 5/16", 3/8", and ½" with its interchangeable drums. This drum machine is our **money maker**. It's used for unclogging kitchen sinks, bathroom sinks, bathtubs, shower stalls, floor drains, yard drains, and even toilet stack lines, when accompanied by the correct drain cleaning attachments. (This is an excellent machine for a property management handyman, drain cleaning professional, and an experienced homeowner.)

Ridgid® K-7500 Drum Machine – which is excellent for cleaning 4" - 10" lines, and can operate four different-sized cables; 5/8", 9/16", 11/16", or ¾" with its interchangeable extra drum option. I prefer keeping two interchangeable drums on

◄ DRAIN CLEANING 101

each service vehicle—one with a 5/8" cable and the other with a 9/16" cable, both 100' in length. I use the 9/16" cable on an everyday basis when unclogging common sewer line stoppages. This cable is more flexible and makes going around bends much easier. I use the 5/8" cable when working on a much tougher blockage such as roots in a sewer line, where a more rigid cable is needed. (This is an excellent machine for a property management handyman or drain cleaning professional.)

The K-3800 and K-7500, if purchased new and properly maintained and cared for, will last you a very long time, and rest assured, you will get your money's worth. I personally use these two machines in my business, and they continue to operate as smoothly as the day I bought them.

Two other machines that I would like to introduce are also by Ridgid®: the **K-40 sink machine** and the **Auto-Clean™ K-30 Sink Machine.** Although I don't use them often, I find them helpful when working in very tight quarters. These two machines are an excellent choice for the **do-it-yourself** homeowner or handyman who doesn't come across many clogged drains. They are small, compact, easy to store away, and they cost much less than the professional models that I use on a daily basis (K-3800, K-7500).

The **K-40 sink machine** is great for cleaning small, secondary drain lines from 3/4" to 2 1/2" such as bathroom sinks, bathtubs, showers, laundry lines, and even the kitchen sink, and uses either a 5/16" or a ¼" cable.

TYPES OF DRAIN CLEANING MACHINES

The **Auto-Clean™ K-30 Sink Machine** is great for cleaning small secondary drain lines from 3/4" to 1 1/2" such as bathroom sinks, bathtubs, and showers, and uses a ¼" cable. (Perfect machine for the occasional clog and is a good choice for homeowners.)

Drum Machine vs. Hand Drill Snake

What's the difference, since they are both designed to unclog common bathtub and sink line stoppages?

In a professional drain cleaner's point of view, there is a big difference, and that would be the ability to work most effectively and safely. I prefer the drum machine over the hand drill snake for a couple of reasons:

1. The drum machine can hold more cable, and this will ensure that you reach and pass the blockage on the first attempt, allowing you to remove most of the blockage from the drain line.
2. Then there are safety reasons. The drum machine is a much safer tool to work with; it stands free, allowing you to position the drum to your desired work angle (if using the K-3800), and it allows you to keep both hands on the cable when in use—it is very important to have control of the drain cleaning cable.
3. The hand drill snake is not my preferred tool, since it is very difficult to operate by a single person. Doing so may result in injury. With the hand drill snake you don't have much control over the cable and the spinning drum when in use.

DRUM MACHINE VS. HAND DRILL SNAKE

4. Most professional drain cleaners rest the hand drill snake machine on their thigh when in operation mode, which is not a very safe practice as the machine can generate friction that may result in injury.
5. In my business, all my work is guaranteed, and since I only use professional drum machines, I can rest assured that when I give a guarantee, I don't have to worry about a recall or comeback. If I used a hand drill snake, I would think twice about giving a full guarantee on my work, and the job would always sit in the back of my mind—I would wonder if the job was efficiently performed.
6. Finally, the last reason why I prefer using a drum machine over a hand drill snake is professionalism and cleanliness. The drum machine is the more professional tool for the job. It gets the job done right the first time, with very little or no mess unlike the hand drill snake.

As a professional you want to uphold your professional image by using professional equipment.

Presentation is very important, both to you and your customer. In my opinion when a client hires a drain cleaning contractor they expect the job to be performed in the most professional manner possible, as they are paying good money for the service.

Showing up with a hand-drill snake in a bucket with a couple of hand tools, as I have seen many do, not acceptable in my eyes. Showing up with a Ridgid® K-3800 Drum Machine, now we're talking. It's the more professional tool for the job, and satisfies your customer's concerns every time.

DRAIN CLEANING 101

Honestly, if you rely on drain cleaning to be your main source of income; do you want your reputation in the business to be a professional one, or to be recognized as some guy with a bucket of tools and a hand drill snake? C'mon...

Basic Use of Your Drain Cleaning Machine

I RECOMMEND THAT when purchasing a drum machine, that you thoroughly read the operating instructions it comes with. This will help you understand its operation completely. It is not a toy; it is a professional piece of equipment and should be treated as such.

When using your drain cleaning machine to unclog blockages, always remember not to pull too much cable out of the drum at one time when feeding it into a drain line. You want to keep control over your cable by keeping a short distance between the drum machine and the entrance of the pipeline you are working on. It's best to keep the cable as straight and firm as possible, as this will help prevent injury and cable kinking. Always operate the machine in the forward position.

For more information regarding proper techniques on drain cleaning cable handling, visit our video section at: www.marinisseweranddrain.com

Drain Cleaning Technique Comparison:

To help explain the basic drain cleaning cable handling technique, what can we compare it to?

Give up??? FISHING!

Let me explain...

Drain cleaning is like fishing. If you have ever gone fishing, you will understand how I came up with this comparison, which is focused on *touch*. When you're fishing, you wait to feel the fish nibble on your line. In drain cleaning, you can feel the blockage when your cable touches it (or nibbles at it). This is why it is very important to concentrate and pay attention to what you're doing.

Along with focusing on touch, listen carefully to the motor on your drum machine. It may start to wind down a bit because it is building tension due to your cable snagging the blockage. The object is not to allow the machine motor to wind down, but to feel the cable nibbling at the stoppage while keeping the cable moving freely throughout the line.

DRAIN CLEANING TECHNIQUE COMPARISON

So, when you feel the tension of a fish nibbling on your bait, you snag it onto your hook by steadily pulling up slowly on your fishing pole. In drain cleaning, the machine and cable do all the work in snagging the blockage. All you're doing is preventing the tension from building up by keeping the cable moving freely about the waste line. Keep both hands on the cable and steadily work it slowly in and out of the line, allowing the cable to travel freely throughout the waste line. When you feel that you have cleared and passed the blockage, keep the machine running and slowly and steadily extract the cable from the line and back into the drum.

Hopefully my brief explanation will help you understand why touch and the ability to feel are very important in drain cleaning.

To learn more about my drain cleaning cable handling techniques, visit my website and watch our videos. We demonstrate how to clean every type of stoppage mentioned in this book.

Drainage System Terminology:

In this section you will learn the different terminologies used in explaining how a drainage system operates, and how all fixtures connect within the system and where they drain into.

Let's get started...

Your drainage system consists of the following: plumbing fixtures, drain lines, traps, waste lines, stack lines, branch lines, and a main sewer line.

It all starts with a plumbing fixture (sink, bathtub, shower, and toilet). How do they connect into the drainage system?

1. A sink is connected to drain lines, which are located underneath the sink and connect into a J- or P-shaped fitting called a **trap**. (a.k.a. J-bend or P-trap) The trap connects to a **waste line**, which is the small section of piping after the trap that connects through the wall and into a **stack line**. A stack line is usually 2 to 4 inches in diameter and is located inside the wall in a vertical position; it travels down the structure until it reaches the lowest level, where it will connect either into a **branch line**, which then connects

DRAINAGE SYSTEM TERMINOLOGY

into the **main sewer line**, or the stack line may connect directly into the main sewer line.

2. Although you cannot see their drain lines or waste lines, as they run under the bathroom floor, bathtubs and showers have a trap and connect into a stack line, which will be explained as you read on, and the process is the same: The drain lines connect to a vertical stack, and the stack travels down the structure to its lowest point and connects either into a branch line, which then connects into the main sewer line, or the stack line may connect directly into the main sewer line.

3. A toilet bowl fixture has its own drain line, called a **lead-bend**—a 90-degree angle soft piping made of lead. It's not usually used in new homes, but you may still find them in older homes and some apartment buildings. If you still have a lead-bend, I would recommend replacing it with no-hub / cast-iron piping.

4. The toilet lead bend connects from the **toilet flange** to a vertical stack line. A vertical stack line which a lead bend connects into is usually 4 inches or greater in diameter, which travels down to the lowest level of the structure and connects either into a branch line, which then connects into the main sewer line, or may connect directly into the main sewer line.

If the toilet is already located in the lowest level of the structure, then the toilet lead-bend or piping will connect into a sewer branch line, which connects into the main sewer line or the lead-bend may connect directly into the sewer line.

* **toilet flange** is a thin metallic ring/disc that is secured to the floor under your toilet bowl and has a dual purpose. It secures the lead bend in place under your toilet

and also serves as a bracket that the two toilet bolts (a.k.a. flange bolts) slide into so that the toilet can be secured in place.

5. A branch line is a horizontal waste line located on the lowest level of the structure, usually under the floor, that handles all the wastewater that drains from the fixtures and stack lines before it reaches the *main sewer line and main house trap*.

6. What is a main sewer line and how do we locate it? The main sewer line handles the flow of wastewater from the entire structure before it drains out into the city sewer or septic tank.

The main sewer line can be located either in the floor (which you cannot see) or exposed and above ground, both at the lowest level of the structure. As wastewater flows down the drain lines and into the main sewer line, it passes through a trap—the main house trap—before it exits the structure. Once the water passes the trap, it will flow into the city sewer or septic tank. The main house trap is also located on the lowest level of the structure since it is connected to your main sewer line.

You can usually find the main house trap toward the front of your structure—either inside or outside close to the foundation wall. Two of the most common places to check would be a basement or in the garage. For example: In New York, if you can locate your water meter, your main house trap is usually close by. If it is not exposed, it is usually located under an access panel on the floor.

Once located and if it is exposed, the main house trap is a large U-shaped pipe with two brass or lead plugs on top. If it is not exposed and is located in the floor, all you should see are the two caps. Keep in mind that if you never had to locate the main house trap before and it is covered in a small pit, there is the possibility that an accumulation of dirt may have covered the caps. So you will need to search for them.

Another method to locate your main house trap is to locate the vent line. The vent line allows air to circulate through your drainage system to help keep your drain lines working properly. Have you ever noticed a round piece of piping that may be protruding from the front or front-side of your structure? Perhaps there is a **flat-screened cover** on the front wall of your structure, or perhaps you have a piece of piping that resembles a **candy cane**. These two items are the **main sewer line vent**.

Now, once you have located one of these two items, take a mental picture of its location and head into the lowest level of the structure; after you get into the approximate area, you should be able to locate the main house trap, as it should not be that far away from the inside or outside of the foundation wall.

Sometimes you can see the vent line passing through the structure wall, so all you need to do is trace it down toward the floor, and it will lead you right to the main house trap. Look along the floor for a cover. If the sewer line is exposed, then the trap should be exposed, and you may notice the house trap hanging along the foundation wall.

If the main house trap is above ground or hanging, it may be located in an unfinished basement or behind a wall out of sight in a finished basement, garage, or ground-floor living quarters. Hopefully an access panel can be located.

Types of Drain Line Blockages: How to diagnose, unclog, and prevent future blockages

Common household drain line blockages:

    Kitchen Sink
    Bathtub
    Bathroom Sink
    Shower Stall
    Toilet
    Laundry Sink
    Floor Drain
    Yard Drain
    Roof Leader
    Sewer Line

NOTE: In the unclogging of kitchen sinks, bathroom sinks, laundry sinks and utility sinks, you will be inserting your drain cleaning cable into the **clean-out** located on the bottom of the sink trap. REMINDER: A sink trap is the J- or P-shaped metallic or PVC (a.k.a. plastic pipe) plumbing fitting located under your sink that the sink

TYPES OF DRAIN LINE BLOCKAGES

drain lines connect into before passing behind the wall and entering the vertical stack line. Keep in mind if the sink trap does not have a clean-out, you can either remove the trap or make a drill-out. (see page. 74)

***clean-out** is a convenient area to access a drain pipe to clear stoppages.*

SOMETHING TO REMEMBER, WHEN clearing stoppages, especially in vertical stack lines, to assure yourself that the drain cleaning cable is traveling down the line towards the stoppage, instead of up the line and going in the wrong direction, you will notice one of the following;

1. If and when you are feeding a drain cleaning cable into a line, if the cable seems difficult to guide in, or feels loose or light and may be pushing itself out of the line, then most likely, the cable is traveling up the line instead of down.
2. If you are feeding a drain cleaning cable into a line, and you notice that the cable seems to be traveling easily into the line, then most likely the cable is traveling down the line.
3. If you are having difficulty determining which direction your cable is traveling try this; pull back on the cable slightly and if you feel tension, or the cable is a bit difficult to pull back, then rest assured that the cable is traveling in the correct direction down the line towards the blockage.

WHY IS MY KITCHEN SINK CLOGGED?

The kitchen sink usually gets stopped up for a couple of reasons. There might be a washing machine hookup connected directly

◄ DRAIN CLEANING 101

to the drain line, which will eventually cause a lint build-up. The kitchen sink may also get clogged by washing dishes without a sink strainer. Food particles or particles from steel wool or scouring pads may also wash down the drain causing a blockage. Then there is the pouring of grease (after cooking /frying) down the sink, and now—saving the best for last—I wonder what may have been dropped down the drain? Perhaps a Popsicle stick, a pen, or a straw...the possibilities are endless.

HOW TO DIAGNOSE A KITCHEN SINK STOPPAGE:

There are two types of diagnoses: Is it a **local stoppage** or a **stack line stoppage?**

In drain cleaning terminology, a **local stoppage** is a clog in the sink drain line before it reaches the sink trap under the sink (which is the piping between the bottom of the sink strainer and leading into the sink trap). A **stack line stoppage** is a stoppage that is located somewhere down the vertical drainpipe in the wall that the sink line connects into.

HOW DO WE DETERMINE BETWEEN THE TWO?

- If you have standing water in the sink, scoop it into a pail until the water level gets as low to the sink drain as possible. Then discard the water in the pail into the toilet. Don't scoop out all the water; leave some in the drain, as it will help determine what type of stoppage you are up against.
- Take a small pail or container and place it under the sink trap. (I prefer using a reinforced kitchen garbage bag, but if you're a novice, use the pail.) Remove the nut on the bottom of the sink trap slowly allowing any water to drain. This will tell you what type of stoppage you may have.

- Once the sink trap nut is removed, if the water standing in the sink drain is still there, then you have a **local stoppage**, meaning the stoppage is before the sink trap. Your obstruction is between the sink drain and the trap and can be unclogged easily by disconnecting your drain lines, removing the stoppage, and then reassembling the drain lines to their original position.
- Keep in mind that if you remove the trap nut and you notice the trap to be clogged, manually remove the clog, allowing the water in the sink to slowly drain. This may be the only cause of the blockage. Replace the trap nut and run some water to test drainage, and you may be done. (This is a very easy fix.)
- If the water sitting in the sink drain flushes into the pail when the trap nut is removed and the trap appears to be clear of any obstructions, then you have a stack line stoppage, meaning the stoppage is after the trap and somewhere down the vertical waste line in the wall.

HOW TO UNCLOG A KITCHEN SINK STOPPAGE:

As you now know from your diagnosis, if it's a local stoppage (before the sink trap), it's an easy fix. Dismantle the drain lines and make sure to keep a small pail underneath to catch the water from the sink strainer and drain pipe. Then you can remove the blockage, reset your drain lines and don't forget to replace the trap nut.

If the blockage is after the trap and in the vertical stack line, then you need to use your electric drain cleaning drum machine, preferably the Ridgid® K-3800 model with the small sink drum with either the ¼" cable or the standard drum with the 3/8" cable. Both are designed to clear stoppages in the kitchen

DRAIN CLEANING 101

stack line. (Homeowners can use the **K-40 sink drum machine** or the **Auto-Clean™ K-30 Sink Machine** with a ¼" cable.)

Once your drain cleaning machine is set up and ready, manually insert the drain cleaning cable into the waste line through the clean-out at the bottom of the sink trap where you previously removed the trap nut. These cables are quite flexible, so manually push the cable up and through the bottom of the trap into the line as far as it can go, towards **"the drop"** also known as the entrance to the vertical stack line, which is through the wall.

Then place both hands on the cable and apply a little forward pressure so that when you step on the foot pedal and turn on the machine, you can lightly push and guide the drain cleaning cable down the line, GO FISHING, and catch the blockage.

Keep in mind that once you feel you've hit the stoppage, you may need to work the cable in and out a bit to pass and clear the stoppage. Once you pass the blockage, put in an extra two or three feet of cable to ensure the line is clean. Remove the cable, slowly feeding it back into the drum, and when you feel you're about two feet from the sink trap, ease up off the foot pedal, turning off the machine. Now remove the remainder of the cable from the drain line manually. This will help prevent splashing of water or debris that is being removed from the drain line, resulting in an easier clean-up.

At this time your kitchen sink stoppage should be cleared, and you can reinstall the sink trap nut and perform a water test. Make sure that the sink drain lines are secure. Then run the water in the sink for a few seconds—count to ten slowly (one Mississippi, two Mississippi, and so on). If you get to ten and there is no sign of a backup, the drain is working properly.

TYPES OF DRAIN LINE BLOCKAGES

If there is no sign of a backup, place the stopper in the sink drain and fill the sink with lukewarm water approximately two inches from the brim of the sink. Once you have reached this water level, remove the stopper. This will allow the weight of the water in the sink to rapidly drain and flush the drain line of any particles left behind from the cleaning. At this time your sink drain should be draining quickly, and your job is complete.

NOTE: Always keep in mind that when you're feeding the drain cleaning cable into a drain line, **mentally measure** by using your hands as a ruler. Pull approximately one foot at a time out of the drum as you feed it into the drain line. This technique will help you avoid putting in too much cable. All you want to accomplish is to unclog the drain line and remove the blockage.

** Very Important: Never place your drain cleaning cable directly into the drain line through the sink strainer drain. This is not the proper way to clean the line. By doing this, you may cause damage to the drain lines, as they are made from a thin gauge of rough brass and are not designed to handle the pressure of the cable while it is in motion. Always work from under the sink as previously mentioned to help avoid any unnecessary repairs. **

HELPFUL TIPS TO PREVENT A KITCHEN SINK STOPPAGE:

Invest in a stainless steel mesh sink strainer: If kept in your sink drain at all times, it will prevent any small particles from entering your drain lines.

Disconnect your washing machine: Many homeowners have their washing machines hooked up directly to their kitchen sink drain line, which will eventually cause future blockages because the lint is not being trapped and discarded.

To prevent future stoppages caused by washing machines, move your washing machine into the basement or lowest level and have your plumbing professional connect your washing machine discharge directly into your sewer line. The sewer line is much larger in size than your kitchen line and can handle the volume. If you need to keep your washing machine in the kitchen, have the water discharge directly into the sink and either keep the strainer on the drain or use a rubber band to connect a baby sock or a small nylon sock over the end of the discharge hose (let it dangle about three to four inches) and collect the lint.

Also, use liquid laundry detergent rather than powder, as a liquid detergent dissolves, and powder can cake up in your drain line, causing a blockage over time.

DIY Drain Maintenance: As the saying goes; an apple a day keeps the doctor away. Well, preventive maintenance helps keep the plumbing professional away. Maintain your household drains with **BIO-CLEAN**® or **Maximizer DT Pro**™, which are a special combination of natural bacteria and enzymes that DIGEST organic waste found in your plumbing system: grease, hair, soap scum, food particles, paper, cotton...etc.

For our complete line of drain maintenance products, visit us online at: www.marinisseweranddrain.com

Grease Disposal: Let cooking grease cool, then place it into a metal can that can be covered or into a grease disposal container (Fat Trapper). Properly dispose of it into the trash can—not down the drain.

Never Use Harsh Chemicals or Over-the-Counter Drain Openers: You're just wasting your money. In my experience, they hardly work, they're hazardous, and they are not good for your plumbing system. A drain line needs to be electrically cleaned first, and then properly maintained to prevent future blockages.

Do Not Install a Garbage Disposal: I am not a fan of the garbage disposal. Food, peels, and vegetable shavings are not to be ground up and washed down your kitchen sink drain. Over time this will cause your kitchen sink line to back up. Do yourself a favor and install a sink strainer to catch anything that may fall into the sink, and discard into the trash. This will save you money on future visits from your drain cleaning professional and also eliminate that occasional foul odor coming from your sink.

WHY IS MY BATHTUB CLOGGED?

Usually a bathtub stoppage is caused by hair or fats and oils from hair/body cleansing products accumulating in the drain or drain line. Or, if the drain screen is missing, something can be dropped down the drain or possibly dropped down the trip-lever overflow or the standing waste overflow.

First let me take a moment to explain the difference between the trip-lever overflow and the standing waste overflow.

The bathtub trip-lever overflow is usually located directly over the bathtub drain below the spout on the inside wall of the tub, and it has a small lever. If the lever is in the upward position, its job is to hold the water in the bathtub, and when in the downward position, it allows the water to drain.

The standing waste overflow is usually found in older homes and apartment buildings. It is roughly a 17" high chrome-plated tube protruding from the floor normally located approximately two to four inches from the exterior wall of the bathtub, adjacent to the tub drain. It has a small handle to lift the internal stopper. Lifting the handle up allows the water to drain, and when in the downward position, it holds the water in the tub.

The two possible ways that an object can be dropped into these overflows are either by the cap or faceplate being removed and the opening left unattended or by not properly securing them during their initial installment.

HOW TO DIAGNOSE A BATHTUB STOPPAGE:

There are a couple of types of diagnoses; is it a **local stoppage** or a **stack line stoppage?**

In drain cleaning terminology, a **local stoppage** in a bathtub is a clog in the bathtub drain line either before the trap or after the trap before it connects into the stack line in the wall. Usually the bathtub drain line connects directly into the toilet stack line or in most cases the bathroom sink stack line in the wall.

> **NOTE:** We can't see either of these stack lines; as they are vertical lines located inside the wall

So how do we determine if it is a local stoppage or a stack line stoppage?

> **Step 1:** Depending on what style bathtub stopper you are working on, before starting, make sure that the bathtub stopper is in the position that allows the water to drain.

> **Step 2:** If you have standing water in the bathtub, leave it there. If you don't, run some water in the tub until you form a small puddle over the tub drain.

TYPES OF DRAIN LINE BLOCKAGES

Step 3: Fill the bathroom sink with water approximately one inch from the rim, and then let the water drain immediately. Now you need to observe. If you notice the water that is draining from the sink is backing up into the tub, then you do not have a local bathtub stoppage. You have a **stack line stoppage**. Now you will need to work from the bathroom sink trap to clear the clog in the stack line. If the water doesn't back up from the sink to the tub, then go to **Step 4.**

Step 4: Now flush the toilet. If you notice the toilet water is backing up into the bathtub, then you will need to remove the toilet and snake the line, aka the toilet stack line.

Also see: sewer line / branch line stoppage for additional instructions.

Step 5: If you perform Steps 2, 3, and 4 and there are no signs of a stack line backup, then you have a local bathtub stoppage.

HOW TO UNCLOG A BATHTUB STOPPAGE:

Now that you have determined that you have a local stoppage:

Keep in mind the two different types of overflows. You have the trip-lever overflow and the standing waste overflow; both are positioned directly above the bathtub trap, which is where you want to insert your drain cleaning cable. Doing so enables your cable to pass easily around the bathtub trap to properly clean the drain line.

DRAIN CLEANING 101

Removing the faceplate of the trip-lever overflow is easy. Simply unscrew the screw(s), which secure the faceplate to the overflow and gently remove the faceplate and connecting stopper from the overflow tube.

If you have a standing waste overflow, with a pair of 12" straight jaw pliers—loosen the top cap on the overflow and then gently remove the cap along with its internal stopper.

Now you're ready...

Using your electric drain cleaning drum machine, preferably the Ridgid® K-3800 model with the small sink drum and ¼" cable, position the machine and tilt the drum to your desired position. (You may also use the Ridgid® **K-40 sink machine** or the **Auto-Clean™ K-30 Sink Machine**.)

Manually insert the cable into the overflow until the tip of your cable is in the bathtub trap and can't go any further. Always remember to be in control of the cable. Keeping both hands on the cable, apply a little downward pressure so that when you step on the foot pedal to turn on the machine, you can easily guide the cable around the bathtub trap.

Once your cable passes the trap, slowly feed it into the line and remember to pay attention to the standing water in the bathtub; once you notice the water starting to drain, you have cleared the stoppage.

As the water is draining from the bathtub, keep the cable in the line, turn on the water in the bathtub so that while you continue to snake the line and remove the cable, the water will flush

TYPES OF DRAIN LINE BLOCKAGES

the pipe line of any lingering particles while simultaneously cleaning your cable. Not to worry—whatever you snagged in the cleaning process will remain on your cable and be removed from the drain line.

Also keep in mind that there is a very short distance from the tub trap to where the line may connect, so you don't want to feed too much cable into the line. You don't want it to get stuck or possibly push the contents of the bathtub stoppage into the adjoining pipeline, possibly causing an additional stoppage.

> ** Very Important: Never place your drain cleaning cable directly into the bathtub drain, as this is not the proper way to clean the line, plus you may also cause damage to the drain line before it enters the trap. The drain line is made from a thin gauge of rough brass and is not designed to handle the pressure of the cable while it is in motion. Always work from the overflow tube to ensure proper cleaning and avoid unnecessary costly repairs.

NOTE: If you are working on a local bathtub stoppage and have snaked the line from the trip lever overflow or standing waste and the water in the bathtub is not draining, it is possible that there may be some hair accumulation in the bathtub drain before the trap. If there is a screen over the drain remove it by loosening the screw in the center. Insert the drain cleaning cable (preferably the ¼" cable) into the drain towards the bathtub trap (usually not more than twelve inches). Gently tap the foot pedal on your drain cleaning machine one time, just enough to spin the cable. As it slowly spins retract it from the drain.

Doing this should remove any hair accumulation that may be clogging the drain line before the trap (this is the only time I would recommend inserting a drain cleaning cable directly into a bathtub drain).

NOTE: When you are feeding the drain cleaning cable into a drain line, **mentally measure** by using your hands as a ruler. Pull out approximately one foot at a time from the drum as you feed it into the drain line. This technique will help you avoid putting in too much cable. All you want to accomplish is to unclog the drain line and remove the blockage.

For a bathtub line you want to mentally measure the distance from the bathtub overflow to where the drain line may connect into, probably into the bathroom sink stack or the toilet stack. In my experience, when unclogging a bathtub stoppage, approximately six to fifteen feet of cable is more than enough, depending on the distance of the tie-in point of the stack line.

HELPFUL TIPS TO PREVENT A BATHTUB STOPPAGE:

Invest in a stainless steel mesh bathtub strainer: If kept in your bathtub drain at all times, it will prevent any small particles or hair from entering your drain lines.

DIY Drain Maintenance: Preventive maintenance is very important to keep your drain lines flowing properly. Maintain your household drains with **BIO-CLEAN®** or **Maximizer DT Pro™**—which are a special combination of natural bacteria and enzymes that DIGEST organic waste found in your plumbing system: grease, hair, soap scum, food particles, paper, cotton...etc.

Never Use Harsh Chemicals or Over-the-Counter Drain Openers: You're just wasting your money. They hardly work, they're hazardous, and they are not good for your plumbing system. A drain line needs to be electrically cleaned first, and then properly maintained to prevent future blockages.

WHY IS MY BATHROOM SINK CLOGGED?

The bathroom sink usually gets clogged due to not having a sink strainer, which can help prevent hair or any object from accidently falling down the drain, such as jewelry, a toothbrush, or other small items.

HOW TO DIAGNOSE A BATHROOM SINK STOPPAGE:

The bathroom sink stoppage and the kitchen sink stoppage are very similar.

There are two types of diagnoses: Is it a **local stoppage** or a **stack line stoppage**?

In drain cleaning terminology, a **local stoppage** is a clog in the sink drain line before it reaches the sink trap (which is the piping located under the sink between the sink drain, and leading into the sink P-Trap or J-Bend). A **stack line stoppage** is a stoppage that is located somewhere down the vertical drainage pipe in the wall that the sink line connects into.

Now how do we determine between the two?

If you have standing water in the sink, scoop it into a pail until you get as low to the sink drain as possible. Then discard the water in the pail into the toilet. Don't scoop out all the water; leave some in the drain, as it will help determine what type of stoppage you are up against.

Take a small pail or container and place it under the sink trap. (I prefer using a reinforced kitchen garbage bag, but if you're a novice, use the pail.) Remove the nut on the bottom of the sink trap slowly allowing any water to drain. This will tell you what type of stoppage you may have.

Once the sink trap nut is removed, if the water standing in the sink drain is still there, then you have a **local stoppage**, meaning the stoppage is before the sink trap. Your obstruction is between the sink drain and the trap and can be unclogged easily by disconnecting your drain lines, removing the stoppage, and then reconnecting the drain lines.

Keep in mind that if you remove the trap nut and you notice the trap to be clogged, manually remove the clog, allowing the water in the sink to slowly drain. This may be the only cause of the blockage. Replace the trap nut and run some water to test drainage, and you may be done. (This is a very easy fix.)

If the water in the sink drains completely when the trap nut is removed and the trap appears to be clear, then you have a **stack line stoppage**. The stoppage is most likely after the trap and somewhere inside the vertical line in the wall.

HOW TO UNCLOG A BATHROOM SINK STOPPAGE:
(Very Similar to a Kitchen Sink Stoppage)

As we know from when we diagnosed our problem, if it's a local stoppage (before the sink trap) it's an easy fix. Just make sure when you dismantle your bathroom sink drain line to keep a small pail under the drain lines you are removing to catch any water that may be in them. Once removed, then you can clear the blockage and reset the drain line(s).

TYPES OF DRAIN LINE BLOCKAGES

Keep in mind that you may also have an accumulation of hair around the bathroom sink stopper. The stopper can be easily removed by loosening the pop-up-assembly nut, which is located directly underneath the sink. On the back side of the drain line close to the top, you will feel or see a small nut that secures the pop-up assembly rod (or stopper) to the inside of the sink drain. Simply loosen and remove this nut (unscrew toward your right), then remove the nut and the stopper rod just enough so that you can remove the stopper from the sink to clean it, then replace. To tighten the pop-up assembly nut, turn the nut to your left. Do not over tighten; make it snug, as tightening it too much will keep the pop-up from working properly.

If the blockage is after the trap and in the stack line, then you need to use your electric drain cleaning drum machine, preferably the Ridgid® K-3800 with the small sink drum and ¼" cable. You may also use the **K-40 sink machine** or the **Auto-Clean™ K-30 Sink Machine,** which are also designed to clear stoppages in a bathroom sink line.

Once your drain cleaning machine is set up and ready, manually insert the drain cleaning cable into the waste line through the clean-out at the bottom of the sink trap where you previously removed the trap nut. This cable is quite flexible, so manually push the cable up and through the bottom of the trap into the line as far as it can go. Then place both hands on the cable and apply a little forward pressure so that when you step on the foot pedal and turn on the machine, you can lightly push and guide the drain cleaning cable down the stack line.

Once you feel you've hit the stoppage, you may need to work the cable in and out a bit to pass and clear the stoppage. After you pass the blockage, put in an extra couple of feet to ensure

DRAIN CLEANING 101

the line is clean. Remove the cable, slowly feeding it back into the drum, and when you feel you're about two or three feet from the sink trap, ease up off the foot pedal, turning off the machine. Now remove the remainder of the cable from the line manually. This will prevent splashing of water or debris from the line, resulting in an easier clean-up.

At this time your bathroom sink stoppage should be cleared, and you can reinstall the sink trap nut and perform a water test. Make sure that the sink drain lines are secure. Run some water in the sink for a few seconds—count to ten slowly (one Mississippi, two Mississippi, and so on). If you get to ten and there is no sign of a backup, the drain is working properly.

If there is no sign of a backup, place the stopper in the sink drain and fill the sink with lukewarm water approximately two inches from the brim of the sink. Once you have reached this water level, release the stopper. This will allow the weight of the water in the sink to rapidly drain and flush the drain line of any particles left behind from the cleaning. At this time your sink should be draining quickly, and your job is complete.

Always keep in mind that when you're feeding the drain cleaning cable into a drain line, **mentally measure** by using your hands as a ruler. Pull approximately one foot at a time out of the drum as you feed it into the drain line. This technique will help you avoid putting in too much cable and cable kinking. All you want to accomplish is to unclog the drain line and remove the blockage. In my experience, when unclogging a bathroom sink stack line stoppage, nine to eighteen feet of cable is more than enough.

TYPES OF DRAIN LINE BLOCKAGES

** Very Important: Never place your drain cleaning cable directly into the drain line through the sink strainer drain, as this is not the proper way to clean the line, and you may also cause damage to the drain lines. They are made from a thin gauge of rough brass and are not designed to handle the pressure of the cable while it is in motion. Always work from under the sink as previously mentioned to help avoid any unnecessary repairs. ******

HELPFUL TIPS TO PREVENT A BATHROOM SINK STOPPAGE:

Invest in a stainless steel mesh sink strainer: If your bathroom sink does not have an internal stopper / pop-up-assembly, it would be wise to invest in a stainless steel sink strainer. This will prevent hair or other small particles from entering your drain lines.

DIY Drain Maintenance: Maintain your household drains with **BIO-CLEAN®** or **Maximizer DT Pro™**, which are a special combination of natural bacteria and enzymes that DIGEST organic waste found in your plumbing system: grease, hair, soap scum, food particles, paper, cotton...

Install a Catch-All Drain Trap: Which easily installs under your bathroom sink enabling you to catch and retrieve anything that may accidentally drop down the drain.

Never Use Harsh Chemicals or Over-the-Counter Drain Openers: You're just wasting your money. They hardly work, they're hazardous, and they are not good for your plumbing system. A drain line needs to be electrically cleaned first, and then properly maintained to prevent future blockages.

WHY IS MY SHOWER STALL CLOGGED?

A shower stall stoppage usually is caused by hair or fats and oils from hair/body cleansing products accumulating in the drain line. Or, by any small object that may have accidentally washed down the drain due to a missing drain cover/screen. A shower stall stoppage is quite similar to a bathtub stoppage, with this difference: Unlike the bathtub, the shower stall does not have a stopper or an overflow.

HOW TO DIAGNOSE A SHOWER STALL STOPPAGE:

There are a couple of types of diagnoses; is it a **local stoppage** or a **stack line stoppage?**

In drain cleaning terminology, a **local stoppage** in a shower stall is a clog in the shower stall drain line either in the trap or after the trap before it connects into the stack line in the wall. The shower stall drain line may sometimes connect directly into the toilet stack line, but most commonly it will connect into the bathroom sink stack line.

> **NOTE:** We can't see either of these stack lines; as they are vertical lines located inside the wall. The shower stall waste line is very similar to the bathtub waste line.

SO HOW DO WE DETERMINE IF IT IS A LOCAL STOPPAGE OR A STACK LINE STOPPAGE?

Step 1: If you have standing water in the shower stall, leave it alone. If you don't, run some water in the shower to fill the drain, as this will help you in determining the type of stoppage.

Step 2: Fill the bathroom sink with water approximately one inch from the top, then let the water drain immediately. If

TYPES OF DRAIN LINE BLOCKAGES

you notice that the water draining from the sink is backing up into the shower, then you do not have a local shower stall stoppage. You have a **stack line stoppage**. Now you will need to snake the bathroom sink stack line to clear the blockage.

Step 3: Flush the toilet a couple of times, and if you notice the toilet water backing up in the shower, then you will need to remove the toilet and snake the line, aka the toilet stack line.

Also see: branch line and sewer line stoppage for additional instructions.

Step 4: If you performed **Steps 2 and 3** and there are no signs of a backup, then you have a local shower stall stoppage.

HOW TO UNCLOG A SHOWER STALL STOPPAGE:

When cleaning a shower stall drain line, since you can't work from underneath, snake the line directly through the drain by simply removing the strainer from over the drain.

Just like a bathtub overflow, the shower stall drain is directly in line with its trap. Not to worry, as the shower stall trap, like the bathtub trap, is made from a heavier metal (usually brass), which is capable of handling the pressure of a drain cleaning cable while in motion.

Using your electric drain cleaning drum machine, preferably the Ridgid® K-3800 model with the small sink drum and ¼" cable, position the machine and tilt the drum to your desired position. You may also use the **K-40 sink machine** or the **Auto-Clean™ K-30**

◄ DRAIN CLEANING 101

Sink Machine, which are also designed to clear stoppages in shower stall drain lines.

Manually insert the cable into the shower stall drain until the tip of your cable is in its trap and can't go any further. Always remember to be in control of the cable. Keeping both hands on it, apply a little downward pressure onto the cable so that when you step on the foot pedal to turn on the machine, you can easily guide the cable around the trap.

Once your cable passes the trap, slowly feed it into the line and remember to pay attention to the standing water in the shower stall. When you notice the water starting to drain, you have cleared the stoppage. As the water drains, continue to snake the line approximately two extra feet, as this will ensure that you have passed the stoppage and thoroughly cleaned the line.

Now you want to start retracting the cable slowly from the line and place it back into the drum. Once you feel the cable has passed back through the trap, slowly ease up on the foot pedal, turning off the machine then completely remove the remainder of the cable from the line manually. Run the water in the shower to perform a water test. Replace the drain screen/cover over the drain opening, and you are done.

Also keep in mind that there is a very short distance from the shower trap to where the line connects, so you don't want to feed too much cable into the line. You don't want to get the cable stuck or possibly push the contents of the shower stall stoppage into the adjoining pipeline, causing an additional stoppage.

TYPES OF DRAIN LINE BLOCKAGES

NOTE: When you're feeding the drain cleaning cable into a drain line, **mentally measure** by using your hands as a ruler. Pull approximately one foot at a time out of the drum as you feed it into the drain line. This technique will help you avoid putting in too much cable. All you want to accomplish is to unclog the drain line and remove the blockage. In my experience in unclogging a shower stall stoppage, eight to fifteen feet of cable is more than enough.

HELPFUL TIPS TO PREVENT FUTURE SHOWER STALL STOPPAGES:

Invest in a cone hair screen: Which is a screen that is placed into the shower stall drain, underneath the drain cover, which allows the water to drain but catches all small particles and hair.

DIY Drain Maintenance: Preventive maintenance is very important in helping keep your drain lines flowing properly. Maintain your household drains with **BIO-CLEAN®** or **Maximizer DT Pro™**, which are a special combination of natural bacteria and enzymes that DIGEST organic waste found in your plumbing system: grease, hair, soap scum, food particles, paper, cotton...etc.

Never Use Harsh Chemicals or Over-the-Counter Drain Openers: You're just wasting your money. They hardly work, they're hazardous, and they are not good for your plumbing system. A drain line needs to be electrically cleaned first, and then properly maintained to prevent future blockages.

WHY IS MY TOILET CLOGGED?

A toilet bowl stoppage is usually caused by using too much toilet paper, or by using the wrong type of paper product, such as paper towels and baby wipes, as these are never to be flushed down the toilet even if the package says they're flushable. And especially never flush sanitary products.

Another reason for a toilet stoppage may be because someone in the household is on a certain type of medication prescribed by their doctor and one of the side effects may cause hard stool.

Then there is my favorite—someone dropped an object into the toilet. I have removed many interesting items from toilets: cell phones, washcloths, deodorant sticks, perfume bottles, car/house keys, toys, pets, Q-Tips, and my all-time favorite—false teeth… And the list goes on.

HOW TO DIAGNOSE A TOILET STOPPAGE:

A toilet bowl stoppage is quite easy to diagnose since all you have to do is flush the toilet and observe the two possible scenarios of what may happen.

Once again we are determining if it is a **local stoppage** or a **stack line stoppage.**

Now how do we determine between the two?

In drain cleaning terminology, a **local toilet stoppage** is a clog in the toilet bowl, that when flushed, prevents the toilet from draining properly.

If it is a **toilet stack line stoppage**, you may be able to flush the toilet, but it may appear to be draining a bit sluggishly and will most likely back up in the bathtub or shower stall. You may also notice water leaking from around the base of the toilet, then you will have to remove the toilet and snake the stack line. Upon removal of the toilet, you may also notice that the toilet

TYPES OF DRAIN LINE BLOCKAGES

drain line / lead bend may be full of water and not draining. This is very uncommon in single-family homes, but is very likely to occur in apartment buildings.

Also see: How to clear a toilet stack / branch line stoppage, for additional instructions.

Another determination of a toilet stack line stoppage is a scenario that is very common in three-family homes or apartment buildings where there may be other toilets above that share the same waste/stack line. For example, if you have a tenant in an apartment above you, and they flush their toilet and the water drains perfectly, but the toilet in your apartment starts to back up and overflow on its own, then you have a stack line stoppage. Although your toilet may overflow on its own, this may also cause the bathtub or shower stall to back-up as well.

To clear this type of stoppage, you must remove the toilet in the apartment where it is backing up, and snake the toilet stack line.

HOW TO UNCLOG A TOILET STOPPAGE:

Now that we have determined that you are dealing with a local toilet stoppage, the fix is quite simple.

First you may want to try using the toilet plunger. A couple of pumps with the plunger may be all it needs. But if the plunger is not working for you, then you should use a toilet auger as explained below.

This procedure requires the use of the **toilet auger** only. I prefer using the **General T6FL Six-foot Teletube® bulb head toilet auger.**

◀ DRAIN CLEANING 101

Step 1: Depending on which hand is your power hand—meaning if you're a lefty or a righty—you'll want to use your power hand to work the manual crank handle at the top of the auger while your other hand holds the middle grip, keeping the auger steady.

Step 2: Release the snake from the clip on the auger so that it is dangling freely.

Step 3: Grab the center grip on the auger with your desired hand, and with your power hand pull upward on the auger crank handle as far as it will go. At this time you will notice the snake sliding into the auger shaft, bringing the bulb head of the snake close to the bottom part of the auger shaft.

Step 4: Approach the toilet bowl with your auger in hand, and place the bottom part of the auger (bulb side) into the bottom of the inside of the toilet. Position the bulb head so that it is facing and is inserted into the mouth of the toilet, or in the direction that the toilet drains.

Step 5: Now you should have your hands in their correct positions, and remember to keep the auger steady.

Step 6: Using your power hand, start cranking the auger crank handle in whichever direction you're most comfortable (I prefer clockwise), and at the same time as you're cranking, you need to steadily and slowly crank in a downward pushing motion, allowing the snake to travel through the inside of the toilet fixture.

TYPES OF DRAIN LINE BLOCKAGES

Once you get the crank handle all the way down to the top of the auger shaft, then the toilet stoppage should be cleared. All you need to do now is remove the snake from the toilet, steadily and slowly pulling up on the crank handle until the snake is completely removed from the toilet. Job complete.

- Sometimes it takes more than one attempt in clearing a local toilet stoppage with an auger.

Here's another scenario—*a little more time-consuming, but you can do it.*

Now let's say someone accidentally dropped an object into the toilet such as a child's toy (a small action figure), and it accidentally gets flushed and clogs up the toilet. You try your best to remove it from the toilet with your toilet auger, using the step-by-step procedures above, but you have no luck. What do you do now? Let me explain.

You will need to remove the toilet so that you can remove the object, and this is how it's done:

Step 1: Turn the water supply valve off under the toilet, which supplies the water to the toilet tank.

Step 2: Remove the toilet tank cover and place it in a safe location so it doesn't break.

Step 3: Using your Shop-Vac, remove the water from the toilet tank and then the toilet bowl. If you want to try inserting the Shop-Vac hose into the mouth of the toilet to try and

DRAIN CLEANING 101

suck out the object, it's worth a shot—you may get lucky. If that doesn't work, proceed to **Step 4**.

Step 4: Once all the water is removed from the tank and bowl, make sure that the tank is not refilling. This means that the toilet supply valve is not holding, and you may need to shut down the water main.

Step 5: Once you have determined that the water is turned off and the toilet valve is holding, you can loosen and disconnect the toilet water supply line from the bottom of the toilet tank.

Step 6: Loosen and remove the two nuts and washers from the flange bolts that secure the toilet to the floor. Once this is done, you're ready to remove the toilet.

At this time I suggest laying a cloth drop cloth (painter's drop cloth) or even an old blanket or large beach towel down in your work area so that you can place the toilet on top to prevent it from slipping while you're working.

Now you are ready to lift the toilet and position it upside down so that the outlet of the toilet is facing upward in a semi-slanted position. This procedure is best when you have a helper so that they can hold and prevent the toilet from moving when it's flipped over. If you are working alone, as I always do, place the toilet in the bathtub and use the bathtub wall as your helper.

Step 1: Keep the toilet seat in the closed position or remove the toilet seat from the toilet.

TYPES OF DRAIN LINE BLOCKAGES

Step 2: Remove the toilet from its original position by lifting the toilet straight up off the bolts.

Always remember to lift with your legs (bend your knees), as this is a heavy job, and always be careful when maneuvering the toilet, as it is very fragile and can easily crack or break. I recommend wearing a pair of work gloves that fit snugly (for hand protection) so you can grab the toilet and work with ease. **(Don't forget to always wear your safety glasses.)**

Step 3: Gently place the toilet down onto the drop cloth or in the bathtub and maneuver it into its final position, which is the front of the bowl (where you would sit), with the front top section of the tank resting on the floor. This position may be a bit rocky, which is why an assistant is helpful; otherwise, place the toilet into the bathtub so that the toilet can rest against one of the bathtub walls. If you are going to lay the toilet down into the tub, don't forget to put your drop cloth or old blanket down first to protect the bathtub's finish. Now that the toilet is in the correct position, you should be able to see the toilet bowl outlet. (I know seeing the toilet bowl in this position seems odd, but this technique works well.)

Also see: Our video section at www.marinisewageranddrain.com for other toilet bowl maneuver options.

Step 4: Using your flashlight, look inside the toilet outlet. If you can see the obstruction and it's small enough to grab and fit through the outlet, try using needle-nose pliers to remove it. If you can't, go to Step 5.

Step 5: Use your toilet auger in the same manner as you would to clear a toilet stoppage. You may be able to push the stoppage from the toilet outlet through the toilet line and remove it from the toilet bowl.

I use this technique often in removing objects from a toilet, but it is best when performed with two people, so that one person can hold the toilet in place while the other person works the auger.

NOTE: There is always the possibility that the obstruction in the toilet cannot be removed, which will result in replacing the entire toilet bowl fixture which is another easy do-it-yourself project.

Step 6: If the object was removed, you can now reset the toilet and return it to its original working condition. When resetting the toilet, always use a new wax gasket. Doing this will ensure that there is a complete seal between the toilet and the toilet waste line/lead bend, preventing any air gaps while also preventing any water leakage when the toilet is flushed.

Don't forget to re-fasten the toilet to the floor. Do not over tighten the bolts. Snug is just fine. Reset the water supply line and replace the toilet tank cover. Once your toilet is reset, turn the water valve back on slowly, and flush the toilet to check for leaks at the base. If there aren't any, job well done.

NOTE: Never use an electric drain cleaning machine to snake the inside of your toilet bowl.

HELPFUL TIPS TO PREVENT FUTURE TOILET STOPPAGES:

Install a Toilet Seat Lock: Which is a durable, easy to install, effective device that keeps the toilet seat cover down so that it can't be opened by young children.

Keep the Toilet Area Clear: Keep all small decorative items and toiletries from around the toilet, as this will prevent anything from falling in. **Also get into the habit when leaving the bathroom to put the seat and cover down...** *Guys, women will love you for this.*

Do Not Flush These Items: Q-Tips (they don't always go down; they tend to lay across the toilet outlet and eventually cause a stoppage); paper towels (they don't break up); baby wipes (the package says they are flushable, but I wouldn't recommend it.); sanitary products (they don't break up at all and are also the #1 cause for a sewer line backup). All of these items need to be thrown in the trash, not flushed down the toilet bowl.

Do Not Use Hanging Toilet Bowl Deodorizers: They may break off the side of the toilet and mistakenly get flushed. If you want to use a deodorizer, get the type you place in the tank.

WHY IS MY LAUNDRY SINK CLOGGED?

The laundry sink / utility sink / slop sink is a deep sink that is usually located in the laundry room or unfinished section of the basement. This sink is often used for washing machine discharge and rinsing and/or wringing out your string mop or perhaps washing items that you would not usually wash in your kitchen sink.

How do these types of sinks usually get clogged? There may be a washing machine hook-up connected directly to the drain pipe, which will eventually cause a lint buildup within the drain line. Another cause is by not having a sink strainer in the drain, which will prevent strands from a mop or particles from steel wool or scouring pads from washing down the drain. Then there is the possibility that something was accidentally dropped down the drain (a Popsicle stick, a pen, or a straw—the possibilities are endless).

HOW TO DIAGNOSE A LAUNDRY SINK STOPPAGE:
Very similar to the kitchen sink stoppage:

There are two types of diagnoses: Is it a **local stoppage** or a **stack line stoppage**?

In drain cleaning terminology, a **local stoppage** is a clog in the sink drain line before it reaches the sink trap under the sink (which is the piping between the bottom of the sink strainer leading into the sink trap.) A **stack line stoppage** is a stoppage somewhere down the vertical drainpipe in the wall that the sink line connects into.

HOW DO WE DETERMINE BETWEEN THE TWO?
- If you have standing water in the sink, scoop it into a pail until the water level gets as low to the sink drain as possible. Then discard the water in the pail into the toilet. Don't scoop out all the water; leave some in the drain, as it will help determine what type of stoppage you are up against.
- Take a small pail or container and place it under the sink trap. (I prefer using a reinforced kitchen garbage bag, but if you're a novice, use the pail.) Remove the nut on the bottom of the sink trap slowly allowing any water to drain. This will tell you what type of stoppage you may have.

- Once the sink trap nut is removed, if the water standing in the sink drain is still there, then you have a **local stoppage**, meaning the stoppage is before the sink trap. Your obstruction is between the sink drain and the trap and can be unclogged easily by disconnecting your drain lines, removing the stoppage, and then reassembling the drain lines to their original position.
- Keep in mind that if you remove the trap nut and you notice the trap to be clogged, manually remove the clog, allowing the water in the sink to slowly drain. This may be the only cause of the blockage. Replace the trap nut and run some water to test drainage, and you may be done. (This is a very easy fix.)
- If the water sitting in the sink drain flushes into the pail when the trap nut is removed and the trap appears to be clear of any obstructions, then you have a **stack line stoppage**, meaning the stoppage is most likely after the trap and somewhere down the vertical waste line.

HOW TO UNCLOG A LAUNDRY SINK STOPPAGE:

As we know from when we diagnosed our problem, if it's a local stoppage (before the sink trap), it's an easy fix. Just make sure when you dismantle your laundry sink drain lines to keep a small pail underneath to catch any water that may be in them. Once removed, then you can remove the blockage and reset your drain lines. Don't forget to replace the trap nut.

If the blockage is after the trap and in the vertical stack line, then you need to use your electric drain cleaning drum machine, preferably the Ridgid® K-3800 model with the small sink drum and either the ¼" cable or the standard drum with the 3/8" cable. Both are designed to clear stoppages in the laundry

◄ DRAIN CLEANING 101

line. (Homeowners, use the **K-40 sink drum machine** or the **Auto-Clean™ K-30 Sink Machine.**)

Once your drain cleaning machine is set up and ready, manually insert the drain cleaning cable into the waste line through the clean-out at the bottom of the sink trap where you previously removed the trap nut. These cables are quite flexible, so manually push the cable up and through the bottom of the trap into the line as far as it can go. Then place both hands on the cable and apply a little forward pressure so that when you step on the foot pedal and turn on the machine, you can lightly push and guide the drain cleaning cable down the line.

Keep in mind that once you feel you've hit the stoppage, you may need to work the cable in and out a bit to pass and clear the stoppage. Once you pass the blockage, put in approximately two extra feet of cable to ensure the line is clean. Remove the cable, slowly feeding it back into the drum. When you feel you're about two feet from the sink trap, slightly come off the foot pedal, turning off the machine, and remove the remainder of the cable from the line manually. This will prevent splashing of water or debris from the line, resulting in an easier clean-up.

At this time your laundry sink stoppage should be cleared, and you can reinstall the sink trap nut and perform a water test. Make sure that the sink drain lines are secure. Then run the water in the sink for a few seconds—count to ten slowly (one Mississippi, two Mississippi, and so on). If you get to ten and there is no sign of a backup, the drain is working properly.

If there is no sign of a backup, place the stopper in the sink drain and fill the sink with lukewarm water approximately two inches from the brim of the sink. Once you have reached this water level, remove the stopper. This will allow the weight of the water in the sink to rapidly drain and flush the drain line of any particles left behind from the cleaning. At this time your sink should be draining quickly, and your job is complete.

NOTE: Always keep in mind that when you're feeding the drain cleaning cable into a drain line, **mentally measure** by using your hands as a ruler. Pull approximately one foot at a time out of the drum as you feed it into the drain line. This technique will help you avoid putting in too much cable. All you want to accomplish is to unclog the drain line and remove the blockage.

In my experience, when unclogging a laundry sink stoppage, five to fifteen feet of cable is more than enough since the drain line after the laundry sink trap is usually connected to a stack line or branch line that may only be a short distance away.

** Very Important: Never place your drain cleaning cable directly into the drain line through the sink strainer drain. This is not the proper way to clean the line, and you may also cause damage to the drain lines, which are made from a thin gauge of rough brass and are not designed to handle the pressure of the cable while it is in motion. Always work from under the sink as previously mentioned to help avoid any unnecessary repairs. **

HELPFUL TIPS TO PREVENT A LAUNDRY SINK STOPPAGE:

Invest in a stainless steel mesh sink strainer: If kept in your sink drain at all times, it will prevent any small particles from entering your drain lines.

DRAIN CLEANING 101

Disconnect your washing machine and install a lint trap: Although many homeowners have their washing machine hooked up directly to their laundry sink drain line, this will eventually cause future blockages because the lint is not being trapped and properly discarded.

To prevent future stoppages caused by washing machine discharge, you can have your plumber connect your washing machine discharge directly into a sewer stack/branch line by installing a two-inch standpipe. The sewer line is much larger than your laundry line and can handle the volume.

Another preventive tip is to have the water discharge directly into the laundry sink. Either keep the strainer on the drain or install a lint trap. The homemade version has always worked best for my customers: Use a thick rubber band to connect a baby sock or a small nylon sock over the end of the discharge hose (let the sock dangle three to four inches) to collect the lint.

Also, use a liquid laundry detergent rather than powder. Liquid detergent dissolves, while powder can cake up in your drain line, causing a blockage over time.

DIY Drain Maintenance: Maintain your household drains with **BIO-CLEAN®** or **Maximizer DT Pro™**, which are a special combination of natural bacteria and enzymes that DIGEST organic waste found in your plumbing system: grease, hair, soap scum, food particles, paper, cotton...etc. For more information, visit us online at: www.marinisewveranddrain.com

Never Use Harsh Chemicals or Over-the-Counter Drain Openers: You're just wasting your money. They hardly work, they're hazardous, and they are not good for your plumbing system. A drain line needs to be electrically cleaned first, and then properly maintained to prevent future blockages.

WHY IS MY FLOOR DRAIN or YARD DRAIN CLOGGED?

A **floor drain** is usually found in residential homes in the unfinished section of the basement, often in the boiler room area. Not all homes have them. A floor drain is most likely to be found in commercial buildings, restaurants, and parking garages and usually gets clogged when dirt or small debris particles are washed down the drain over time, causing the line to drain slowly and then eventually clog.

A **yard drain** is located outside, around the perimeter of the structure, and is usually found in the driveway, backyard, and even along the sides of the structure. The yard drain's purpose is to keep rainwater from pooling around the structure. A yard drain will usually get clogged when dirt, debris, and leaves are washed down the drain when it rains, especially when the yard drain screen is not properly secured. Over time this causes the line to drain slowly and then eventually clog.

A yard drain may or may not have a trap, depending on your drainage system. Does your wastewater drain into a septic tank or is it connected to city sewer?

If you are on septic, there is a possibility that your yard drains flow into a drywell (which is a holding tank buried under the ground with gravel and dirt in it to act as a drainage field allowing the rainwater to seep into the earth), which most likely will not have a trap.

◄ DRAIN CLEANING 101

If your drainage system is connected to city sewer, then most likely your yard drain line travels into the structure to a trap, then connects into a sewer branch line or may connect directly into the main sewer line. Or, in newer homes into the storm sewer.

HOW TO DIAGNOSE A FLOOR DRAIN or YARD DRAIN STOPPAGE:

There are two types of diagnoses: Is it a **local stoppage** or a **branch line stoppage?**

Here's something new. What is a branch line? It is very similar to a stack line. A stack line, as we already know, is the drain pipe that runs vertically throughout a structure which the household drains such as sinks, bathtubs, showers, and toilets connect into.

As a stack line travels down the structure to the lowest level, it connects into the structure's sewer *branch line* or main sewer line.

A branch line is a drain line that runs horizontally throughout a structure and is located either above ground, hanging from the ceiling of the lowest level, or below ground in the floor of the lowest level of the structure. It eventually connects into the sewer line.

Now let's get back to the subject of floor and yard drains.

In a floor/yard drain, a **local stoppage** is a clog in the drain line before it reaches its trap or in its trap. You will notice water sitting in or pooling on top of the drain, and not draining. If you notice water backing up from the drain, then you have a **branch line** or **sewer line stoppage**. (Also see: sewer line stoppage for a more detailed explanation on how to clear a branch line stoppage.)

HOW TO UNCLOG A FLOOR DRAIN or YARD DRAIN STOPPAGE:

Use your electric drain cleaning drum machine, preferably the Ridgid® K-3800 model with the standard drum, depending on the size of the pipe opening; if it is a two-inch opening, I would recommend using the 3/8" cable with a chisel or spring blade first. If the drain opening is three or four inches in diameter, I would recommend using the same machine with a ½" cable and either a chisel blade or a half blade. (Remember to keep the blade size approximately ½" smaller than the pipe diameter.)

Clearing a floor drain stoppage is much like clearing a stoppage in a shower stall. You cannot see its trap as it is in the floor, directly under the drain opening.

Manually insert the cable into the drain opening as far as it can go until it comes to a stop. At this point the tip of your cable should be in the trap and unable to go any further. Always remember to be in control of the cable. Keeping both hands on the cable, apply a little downward pressure so that when you step on the foot pedal to turn on the machine, you can easily guide the cable around the trap.

Once your cable passes the trap, slowly feed it into the line and remember to pay attention to the standing water in the drain. When you notice the water starting to drain, you have cleared the stoppage.

DRAIN CLEANING 101

As you notice the water draining, continue to snake the line a couple of extra feet, as this will ensure that you have passed the stoppage and thoroughly cleaned the line. Now you want to start retracting the cable slowly, but before doing so, stop the machine and get your water hose. Insert your water hose slightly into the drain and run the water steadily.

While you are slowly retracting the cable, work it in and out from the drain line as you're placing it back into the drum. This will allow you to clean and flush the line, and at the same time you will be cleaning your cable so it's nice and clean for your next job.

Remember, when you feel the cable approaching the trap, ease your foot off the foot pedal, turning the machine off, and then remove the remainder of the cable manually. Once the cable is completely removed from the line, replace the drain cover. Job complete.

Also keep in mind that you don't want to feed too much cable into the line. You don't want to get the cable stuck or push the contents of the floor drain stoppage into the adjoining pipeline, possibly causing an additional stoppage.

Once the floor drain stoppage is clear, you will notice water sitting at the bottom of the inside of the drain. This is normal as the water sitting in the floor drain trap; is there to prevent any gases from entering the structure from its connection into the sewer line.

When clearing a yard drain stoppage, it is always best to locate the trap and attempt cleaning the drain line from there.

TYPES OF DRAIN LINE BLOCKAGES

But sometimes a yard drain trap is located under a floor without access. Before opening the floor, always make your first attempt from the drain if the yard drain trap cannot be located.

A yard drain trap is not that far from the drain—usually a couple of feet inside the structure from the foundation wall.

For example: If you have a yard drain located right outside your basement door, then its trap should be located directly inside the doorway adjacent to the yard drain. There may be a small cover on the floor. If so, lift the cover and you should see a single cap, probably three inches in diameter. That's the yard drain trap. Clean the line using the same technique as in a floor drain stoppage if you can locate the yard drain trap.

If you can't locate the yard drain trap, try clearing the stoppage by first determining the pipe size, and then determine the correct cable size you will need to use. Insert the cable into the drain opening and feed it in manually as far as it can go. Most likely you will be inside or close to the yard drain trap. Keep both hands on the cable at all times and try to pass the trap and clear the stoppage.

This can be a very difficult task for a beginner and even for a seasoned pro, but it can be accomplished. If you feel you are spending too much time on this project, then do your best to locate the yard drain trap.

NOTE: When working from the yard drain trap, it's always best to clean the line both ways; first by passing the trap and cleaning the line into the structure, then out towards the drain.

HELPFUL TIPS TO PREVENT A FLOOR DRAIN or YARD DRAIN STOPPAGE:

Keep the area clean: Try to keep the area around the drain as clean as possible, free from dirt, debris, or any small particles that can be easily washed down.

Install a floor drain strainer: Which is a cone shaped strainer that when inserted into the floor drain, it captures food, debris and other small particles that would otherwise clog your floor drain.

DIY Drain Maintenance: For floor and yard drains; invest in a three inch or four inch canvas flush bag, and by using the flush bag and your garden hose periodically flush your yard and floor drains to keep them flowing freely.

Never Use Harsh Chemicals or Over-the-Counter Drain Openers: You're just wasting your money. They hardly work, they're hazardous, and they are not good for your plumbing system. A drain line needs to be electrically cleaned first, and then properly maintained to prevent future blockages.

WHY IS MY ROOF LEADER LINE CLOGGED?

A roof leader line can get clogged due to outside dirt, debris, and/or leaves that may wash down the gutter and then into the drain line. Sometimes the drain is left uncovered, and an object may fall into the drain line, such as a ball or particles from the roofing material. It's very important to keep your gutters and roof drains properly covered.

Determine which type of drainage your leader line has. Does your leader line come down from the roof and drain onto the

TYPES OF DRAIN LINE BLOCKAGES

lawn or onto a splash block, which is then pitched away from the structure? Or does the leader line connect to an outside drain line that then enters the structure?

If your leader line enters an outside drain line, it can either drain into a dry well or toward and into a trap located inside the structure, connecting either into the sewer line or in newer structures, into a storm sewer.

HOW TO DIAGNOSE A ROOF LEADER LINE STOPPAGE:

There are two types of diagnoses: Is it a **local stoppage** or a **branch line stoppage**?

In drain cleaning terminology, a leader line **local stoppage** is a clog in the leader line/gutter pipe between the gutter and the leader line trap. A **branch line stoppage** is a clog in the drainage pipe located after the leader line trap (J-bend or U-bend), before it connects into the sewer line.

HOW TO UNCLOG A ROOF LEADER LINE STOPPAGE:

If you are clearing a local stoppage, it would be best to disconnect the lowest section of the leader line at the bottom by removing it from the inlet of the drain pipe, as long as the remainder of the leader line is properly secured to the structure. Do this slowly, as the line may be full of water.

If you are unsure if the leader is secured properly, you can make an inconspicuous access hole (drill-out) into the side of the leader line, close to the bottom where it connects into the inlet of the drain pipe. Use a drill with a 7/8" hole saw so that the line may be electrically cleaned from there and avoid having to climb a ladder.

◄ DRAIN CLEANING 101

Using your electric drain cleaning drum machine, preferably the Ridgid® K-3800 model with the small sink drum and 1/4" cable, gently push the cable up toward the gutter drain to unclog the local stoppage. It is much harder to snake upward, as gravity will tend to force the cable downward, so keeping control of the cable is very important. Keep in mind that the drum holds approximately 35 feet of cable which should be more than enough to clear a common household leader line stoppage.

If the stoppage is not between the gutter and the access hole, you may notice water sitting in the drain pipe, if so then try snaking the line down towards its trap. Before doing this, I would advise changing drums to a 3/8" cable with either a small chisel or spring blade.

Manually insert the cable into the drain opening as far as it can go. Always remember to be in control of the cable. Keeping both hands on the cable, apply a little downward pressure so that when you step on the foot pedal to turn on the machine, you can easily guide the cable down the line toward its trap and attempt to pass the trap.

If your cable passes the trap, slowly feed it into the line and remember to pay attention to the standing water in the drain. When you notice the water starting to drain, you have cleared the stoppage.

As the water drains, continue to snake the line a couple of extra feet, as this will ensure that you have passed the stoppage and thoroughly cleaned the line. Now you want to start retracting the cable slowly from the line and place it back into the

TYPES OF DRAIN LINE BLOCKAGES

drum. When you feel the cable is roughly two feet from the drain opening, ease your foot off the foot pedal, turn off the machine, and remove the remainder of the cable manually. Once the cable is completely removed from the line, run a hose in the drain to flush the line and perform a water test. Cover the hole with a boiler plug or wrap it with a piece of gutter tape.

NOTE: If you cannot clear the stoppage and are having a difficult time passing the leader line trap from the drain itself, you will need to locate the leader line trap, which is located on the lowest level of the structure, usually in a basement or crawl space. Clean the drain line from there. Once you locate the trap, keep in mind that since there is a blockage the line may be full of water. I suggest placing a pail under the trap (if possible) to catch any water that may spill out when opened. Remove the trap plug and insert your drain cleaning cable into the trap opening and guide it into the bottom of trap until it can't go any further.

At this time place a little downward pressure on the cable, then step on the foot pedal and guide the cable around the trap cleaning the trap and drain line leading into the connecting branch line or sewer line. Once you have cleared the stoppage extract the cable from the drain line. Now using your flashlight, look into the trap. You will notice an opening on the inside wall of the trap which leads towards the leader line. Once again insert your drain cleaning cable into the trap and guide it into this opening and clean the line towards the leader line. Cleaning this section of the line can also be performed by cleaning the line through the inlet of the drain pipe on the outside wall of the structure and cleaning the line from the drain line inlet into the

DRAIN CLEANING 101

trap. This task is performed by using the 3/8" cable with either a small chisel blade or small spring blade.

When finished don't forget to replace the plug(s) on the trap before performing a water test. Also keep in mind that the roof line trap can be a single or double hole trap. If it is double, cleaning the line is much easier as you would not have to pass the trap to enter the drain line.

This procedure is very similar to a yard drain stoppage when working from its trap.

WHAT IF MY LEADER LINE DRAINS INTO A DRYWELL?

If your leader line drains into a drywell, it does not have a trap. If clogged it is usually because of debris such as; roots, dirt, rocks or leaves.

When clearing this type of stoppage, don't waste too much time, as it can be very difficult—your drain cleaning cable can get easily stuck, especially when entangled in roots.

Is the drain holding water? If not, fill the drain until the water pools a little bit above the drain opening. If the drain is draining slowly, this is a good sign that it is partially clogged and may be easily cleaned.

Use your electric drain cleaning drum machine, preferably the Ridgid® K-3800 model with the standard drum. I would recommend using the ½" cable, which is a firm cable that will clean drain lines up to four inches in diameter, and use a small 1" or 1 ½" single blade first before trying a double blade. When

TYPES OF DRAIN LINE BLOCKAGES

clearing stoppages with blades, it's very wise to start with a small single blade and work your way up to larger-sized blades.

For example, if you are working on unclogging a line that drains into a drywell, by using a small 1" or 1 ½" single blade, you will have a slighter risk of the cable getting stuck especially if their are roots in the line. Also you may be able to bore a small hole through the stoppage, just enough to clear the line. Then you can remove your cable and change the blade to a larger size for a better cleaning.

NOTE: Do not use a blade that is bigger than the pipe size. (For example, if the pipe opening is two inches, then your blade size should not exceed two inches.) My personal suggestion is to always keep the blade size at least ½" less than the actual pipe size that you are working on.

Insert the cable into the line from the drain opening and start your machine. Slowly guide the cable down the line, keeping in mind that with the fishing technique, it is very easy to get snagged on roots. Also be aware that the entrance to the drywell may not be far from the drain opening, so go in slowly and proceed with caution.

If you spend more than one hour trying to clear this type of stoppage, you are wasting your time. It is possible that the drywell needs to be dug up and replaced. I recommend calling your local drain cleaning or plumbing provider to help you determine the best possible solution.

Drywells are common in the suburbs, outside the city limits, and we will not touch much on this subject, as this book is designed to diagnose and unclog household drain lines for homeowners, apartment buildings, and commercial establishments within the city limits.

HELPFUL TIPS TO PREVENT A ROOF LEADER LINE STOPPAGE:

Keep the area clean: Install gutter guards to prevent leaves and debris from washing down the roof leaders.

WHY IS MY SEWER LINE CLOGGED?

A sewer line can get clogged by tree roots, grease buildup, or by flushing paper towels and sanitary products down a toilet.

WHAT IS A MAIN SEWER LINE & WHERE IS IT LOCATED?

Before we continue, let's review what we've learned about main sewer lines.

The main sewer line handles the flow of wastewater from the entire structure before it drains out into the city sewer or septic tank.

The main sewer line can be located either in the floor (where only its trap may be seen) or exposed, above ground hanging from the ceiling or along the foundation wall in the lowest level of the structure.

As wastewater flows to exit the structure, it passes through a trap on the sewer line, which is called the main house trap. This is the final destination on the sewer line that the wastewater passes through before exiting the structure and draining into the city sewer or septic tank.

TYPES OF DRAIN LINE BLOCKAGES

You'll most commonly find the main house trap toward the front of the structure, either inside or outside close to the foundation wall. Two of the most common places to check would be a basement or in the garage. (In New York, if you can locate your water meter, your main house trap is usually close by.) If it is not exposed, it is usually located under an access panel on the floor of the lowest level of the structure.

If it is exposed, the main house trap is a large U-shaped pipe with two brass or lead plugs on top that can be 3 1/2 to 6 inches in diameter. If it is not exposed and is in the floor, all you should see are the two caps. If you never had to locate the main house trap before and it is covered in a small pit, an accumulation of dirt may have covered the caps, and you will need to search for them.

Another method in locating your main house trap is to locate the vent line. The vent line allows air to circulate through your drainage system to help keep your drain lines working properly. Have you ever noticed a piece of piping that may protrude from the front or front side of your structure? Perhaps there is a **flat-screened cover** on the front wall of your structure or you have a piece of piping that resembles a **candy cane**. These two items are the main sewer line vent.

Now, once you have located one of these two items, take a mental picture of its location and head into the lowest level of the structure. This is where you should be able to locate the main house trap, as it should not be that far away from the inside or outside of the foundation wall.

Sometimes you can see the vent line passing through the structure wall, so all you need to do is trace it down toward the floor and it will lead you right to the main house trap. Look along the floor for a cover. If the trap is exposed, you may just see the house trap hanging along the foundation wall.

LINES THAT CONNECT INTO THE SEWER LINE:

Although all wastewater from the drain lines eventually feed into the main sewer line, only two lines actually connect into the main sewer line: stack lines and branch lines.

I have touched on these explanations before, but they bear repeating.

Stack Line: As we already know, secondary lines such as kitchen sinks, bathroom sinks, shower stalls, bathtubs, and toilets all drain into vertical stack lines. These lines are located behind the wall and travel down the structure to its lowest point. Then they connect either into a branch line, which then connects into the main sewer line, or may just connect directly into the main sewer line.

Branch Line: A branch line is a horizontal waste line located on the lowest level of the structure, usually under the floor or exposed and hanging from the ceiling, it may also run along the foundation wall on the inside of the structure. The branch line handles all the wastewater that drains from the structure's stack lines before it reaches and drains into the *main sewer line*.

A toilet on the lowest level can also drain directly into a branch line. In drain cleaning terminology this is called a toilet branch line.

HOW TO DIAGNOSE AND UNCLOG A SEWER LINE STOPPAGE:

If you are a homeowner or have never done this before, I would advise calling your professional drain cleaning provider, as this task requires a heavier machine and professional experience. But by reading this section you will learn how it's done.

Diagnosing and unclogging a sewer line is one of the easiest of all the stoppages. Although it may be the biggest in pipe size and may be a bit messier to deal with, it is the easiest to diagnose and unclog.

A main sewer line clog can be easily determined, but it takes a while to thoroughly explain.

MAIN SEWER LINE STOPPAGE:

The main sewer line and house trap are located on the lowest level of the structure. The six most noticeable places that you may suspect a sewer line backup are through a floor drain, yard drain, toilet bowl, shower stall, bathtub, and utility sink. You will notice that every time a toilet is flushed, dirty water, along with toilet paper particles and feces, will show up in these locations. One other place where you may see water backing up from is the sewer line vent (candy cane or screened cover) located at the outside front of the structure.

Determine if your main sewer line stoppage is a trap-in or a trap-out stoppage or a branch line stoppage:

For the more experienced individual, it is always good to take a preventative measure by checking the city sewer to make sure

DRAIN CLEANING 101

it is not clogged and holding water. This may be classified as a town issue, and can be resolved in the five boroughs of New York by calling the Department of Environmental Protection at 311. If the structure is on a septic system, it is always wise to locate and check the septic tank to see if it is full. (In this case it needs to be pumped out by a licensed pumper.) A septic tank should also be maintained and pumped on a regular basis, preferably once a year.

Now, if neither the city sewer nor septic tank is clogged, we can proceed.

Make sure no one is using water in the home, as any water usage will add to the sewer line stoppage.

Locate the main house trap and slowly remove the plug/cap located near the foundation wall (this is usually **the trap-out side of the main house trap,** which flows toward and into the city sewer or septic tank). Opening this side first is your best first choice, as you may only have a stoppage in the main house trap, which is easy to unclog.

If you notice a little water spraying from under the cap, not to worry—the line may be under a little pressure, so let the water spray until it stops, which means the pressure is relieved and you won't make a big mess when removing the cap entirely.

When removing the cap, if you notice the trap is full of water and not flowing out to the septic tank or city sewer, then you have a trap-out stoppage. A trap-out stoppage is a main sewer line stoppage that is between the house trap and either the septic tank or city sewer.

TYPES OF DRAIN LINE BLOCKAGES

I would not recommend that this task be performed by the average homeowner. It should be handled by a more experienced individual or professional drain cleaner.

TRAP-OUT STOPPAGE:

If your sewer line has a **trap-out stoppage**, when you slowly remove the trap-out plug, the trap will be full of water.

Keep in mind to take mental measurements of the distance of the city sewer or septic tank and where they are positioned on the drainage system, as you don't want to get your drain cleaning cable tangled up in these. Removal can result in replacement of the sewer line if you're not careful. Also look for trees and their location in the front of the structure, as the cause of the stoppage may be from tree roots entering the sewer line.

Now you are ready to clear the stoppage.

Using your Ridgid® K-7500 machine with the 5/8" or 9/16" cable and a 3" single half blade, insert the cable into the sewer line, **trap-out** side of the trap toward the city sewer or septic tank.

Using the same drain cleaning techniques as mentioned in previous stoppages, once you clear the stoppage and notice the water draining from the main house trap, have someone turn on the bathtub on the lowest level or flush the toilet a couple of times to help flush the line as you are clearing the stoppage.

Remember to pass the stoppage just a bit and then start to retract the cable slowly from the line. You have just cleared a main sewer line stoppage (trap-out). If you clean one side of the

◄ DRAIN CLEANING 101

main house trap, you should always clean the other side (**trap-in** toward the structure) as well.

NOTE: If you feel you are hitting a stoppage trap-out and it is tree roots, there is a possibility that it will not clear on the first attempt; it may take a few attempts, especially if the tree is very large.

Also, if you feel that you are having a hard time mechanically cleaning the line or you are getting caught up in roots, retract the cable and call your drain cleaning professional. The line may need to be cleaned using a high-velocity water jet, which cleans sewer lines of roots and hard grease by introducing water into the sewer line under various amounts of pressure ranging from 2000-4000 psi, safely cutting through even the toughest blockages.

If you remove the trap-out cap/plug from the trap, and you notice the **trap-out** side of the main house trap is not flowing through, I would still take the following precautions before attempting to open the trap-in side of the trap.

This is a great tip for the homeowner. It can save you a lot of money by not having to call your plumbing / drain cleaning professional.

I recommend swishing and scraping the inside walls of the house trap with a broomstick, making sure you can touch the bottom of the trap. You may not see it, but there can be a stoppage lurking at the bottom of the trap, and by doing this you may clear the sewer line stoppage.

You may also clean the trap by using your K-3800 drum machine with a ½" cable and a single blade (just as if you were attempting to clear a floor drain stoppage). Snake the main house trap just enough to clear the trap only; doing this will ensure that the house trap is clear and any water pressure that may be on the trap-in side will be relieved and flow through the trap and into the city sewer or septic tank, clearing the sewer line stoppage.

Once you have confirmed that the trap is clear, you can remove the cap from the other side of the main house trap (which is the **trap-in** side) close to the inside of the structure. When this side is open, have someone flush a toilet on the lowest level of the structure with some toilet paper, while you stay by the house trap and pay close attention to the inside of the trap.

After the toilet is flushed, watch the water flow in the main house trap. Wait a bit and you should see the water and toilet paper arrive in the main house trap momentarily. If you don't see any water flowing, then you have a **trap-in** main sewer line stoppage.

What is a **trap-in stoppage**? A trap-in stoppage is a stoppage in the main sewer line anywhere from the main house trap back to the rear of the structure.

Unclogging a Trap-In Stoppage: If you are a more experienced individual or drain cleaning professional, use your Ridgid® K-7500 machine with the 9/16" cable and a 3" single half blade. Insert the cable into the sewer line, trap-in side of the trap, toward the structure. A Ridgid® K-3800 with a ½" cable can also be used for trap-in cleanings, preferably on short runs less than fifteen feet.

DRAIN CLEANING 101

Keep in mind to take mental measurements of the closest stack line, branch lines, toilet, or even floor drains and where they are positioned on the drainage system. You do not want to try and pass them with your cable, as you may go up the branch line or stack line and get stuck. (So sometimes it is best to call your plumbing or drain cleaning professional for all your main sewer line stoppages).

Using the same drain cleaning techniques as mentioned in previous stoppages, once you have cleared the stoppage and notice the water draining through the main house trap, pass the stoppage just a bit and then start to retract the cable from the line slowly. You have just cleared a main sewer line stoppage (trap-in). I also recommend if you clean one side of the main house trap, you should always clean the other side as well (**trap-out** to the city sewer or septic).

- Always remember to reset the caps or plugs to your main house trap securely. If they are not secure, I recommend replacing them with a **lead fit-all plug**, making an airtight seal that will prevent any water leakage or odors from entering your home.

  * **lead fit-all plug** is cast iron cored with lead threads for an easy fit. Although it has threads, it is not to be secured with a wrench. It is to be placed over the clean-out opening and tapped on its center square nut with a lump hammer. Doing this squeezes the lead threads into the clean-out opening, making a water tight seal. To remove the plug tap the sides of the center nut and the plug should loosen.

TYPES OF DRAIN LINE BLOCKAGES

Branch Line Stoppage: When you snake the main sewer line trap-in and you don't clear the stoppage, then it is possible that you have a branch line stoppage. The stoppage may even be located at the connection where the branch line meets the main sewer line, but you are just not catching it by working from the house trap.

The best method to clear this type of stoppage is by removing the toilet on the lowest level if applicable, or through a clean-out on a rear stack line. Using your K-7500 drum machine with the 9/16" cable and a single blade ranging anywhere from 1 ½" thru 3", manually insert the cable into the drain line. If through a toilet line, make sure you pass the lead bend before turning on your drum machine.

On a basement level, you may have a cast-iron sweep rather than a lead bend, but it's always good to make sure to manually pass the lead bend, which will eliminate the possibility of cutting it with the blade.

Using the same drain cleaning techniques as mentioned in previous stoppages, guide the cable down the toilet branch line or rear stack line until it reaches the main house trap or until the water drains. When retracting the cable from the line, remember to shut the machine off when you feel the cable is approximately three feet from the lead bend or stack line opening to avoid cutting the lead bend or causing unnecessary splashing of wastewater from the line.

Once you are done and have cleared the stoppage, you can reset the toilet or recap the stack line.

DRAIN CLEANING 101

HELPFUL TIPS TO PREVENT A SEWER LINE STOPPAGE:

Prevent flushing the following items: paper towels, diapers, sanitary products, baby wipes, cleansing wipes, (if the package says it's flushable, it really isn't). These products don't break up or dissolve the way toilet paper does. Grease from cooking is another main cause of sewer line backups. Grease should be placed in a Fat-Trapper and then properly discarded in the trash.

DIY Drain Maintenance: Preventive maintenance is very important in keeping your drain lines flowing properly. Maintain your household drains with **BIO-CLEAN®** or **Maximizer DT Pro™**, which are a special combination of natural bacteria and enzymes that DIGEST organic waste found in your plumbing system: grease, hair, soap scum, food particles, paper, cotton... etc. These products will help to reduce any grease buildup from forming in the sewer line.

Use RootX®: This is the original, simple, and fast non-metam sodium foaming root control in easy-to-apply powdered formulation. It foams on contact with water to kill roots, inhibit regrowth, and restore pipe flow capacity.

Combination Stack Lines, aka Combo Stack Line:

A COMBO STACK line is a drainage line or stack line in which two or more secondary lines, such as a kitchen sink, bathroom sink, bathtub, and/or shower stall, may connect (aka, a back to back). Combo stack lines are usually found in apartment buildings or any type of structure with multi-family dwellings.

Some examples of a combo:

Back-to-Back Kitchen and Basin Stack Line Stoppage is a stoppage in both the kitchen and basin sink, if they are back to back and share the same vertical stack line in the wall. (i.e., When you run water in either sink, one or the other or both will back up.)

To clear this type of stoppage, you will need to empty both sinks and their drain lines of water. If you are in a multi-family dwelling building, you may notice the water backing up in either sink on its own. Then you will either have to shut the water down or just tell the tenants in the apartments above on that specific line not to use any water. This will help prevent water from backing up as you are working.

◄ DRAIN CLEANING 101

The next step is to determine which sink you will be clearing the stoppage from. First, using your sink machine (I prefer the K-3800 drum machine with a ¼" cable), while working through the sink trap plug, make a mental note of where the stack line may be stationed inside the wall and try to catch the drop into the vertical stack line and clear the stoppage.

If that doesn't work, you may remove the sink drain lines and trap completely, exposing the nipple at the wall / the pipe opening to the stack. Then snake the line directly from there. Also keep in mind that in most buildings, especially older buildings, the sink traps are a heavy gauge brass, either rough brass or chrome plated. If you remove the trap, you may not get it back on, as the galvanized nipple at the wall may be rotten and some of the threads may have worn away and are in need of replacement or repair.

If you notice the trap is a tubular trap, just remove it by unscrewing the nut from the nipple at the wall.

So to avoid unnecessary and costly repairs, you can also do one of my favorites – **The Drill-Out.**

To perform a drill-out you will need the following: a 3/8" corded drill, one #4 countersink drill bit, one 7/8" bi-metal hole saw, and a boiler plug.

Start off using the # 4 countersink bit, which will drill a starter hole through most metals very quickly. Drill a hole in the center of the top bend of the sink trap. (This works best on brass or chrome-plated brass traps.) Then with the 7/8" hole saw, drill the remainder of the hole through the trap. Now you have a fresh clean-out, that's easy to access which allows you to run the cable straight into the stack line.

NOTE: On some older kitchen sink lines, the waste line is galvanized and may make a couple of turns before connecting to the nipple at the wall. In that case, locate the fitting that is connected to the nipple closest to the wall and drill your clean-out hole on the top of that section of pipe. Remember to make sure you have enough room to cover the newly made clean-out hole with a boiler plug.

Now, insert the drain cleaning cable into the new clean-out and clear the stoppage. When finished install and tighten the boiler plug, covering the newly made clean-out. This plug, if installed properly, will not leak and will last a very long time.

Basin and Tub Stack Line – Commonly found in residential homes and apartment buildings, when the bathtub waste line is not an independent line, but connects into the bathroom sink stack line.

Kitchen, Basin, and Tub Stack Line – Also very commonly found in residential homes and apartment buildings, when the kitchen sink and bathroom sink are back to back, and the bathtub connects into that same stack line. In drain cleaning terminology this is known as a three-way or combo stack line.

Diagnosing these stoppages is simple. Many times I have been called to a job and the customer tells me, "My bathtub is stopped up." That is not a lie. Ask the tenant or homeowner a simple question that will help determine a local bathtub stoppage from a basin and tub stack stoppage: "When you run the bathroom sink, does the water back up into the tub"? If the answer is "yes"... then you have a basin and tub stack stoppage. Even though you ask the question, it is always best to test it yourself, as in my experience the customer is usually wrong.

Fill up the bathroom sink (or kitchen sink if you feel it is a three-way) approximately two inches from the brim of the sink and then let the water drain. If you notice the water backing up into the tub, then you have a stack line stoppage.

Here is another way to determine if it is a combo stack line stoppage, especially in a multi-family dwelling: If there are apartments above and someone uses the water in their bathroom sink or bathtub and it backs up into the bathtub on a lower floor in that line, then you have a stack line stoppage.

Toilet and Tub / Shower Combo Stack Line: These are usually located in the same bathroom but can also be back to back with a second bathroom.

Diagnosing this stoppage is easy as well. When dealing with homeowners or multi-family dwellings, your phone call will go a little like this, as the caller is having a frantic fit: **"How soon can you get here? My toilet is overflowing and my bathtub is filling up with !$*% (dirty water)!"**

First you tell the caller that you are on your way. Ask if there are any tenants directly above sharing that same line. If so, ask that they be made aware of the problem and instructed not to use any water in their bathrooms and kitchen sinks (in case the kitchen sink shares the combo stack line) until the situation is resolved. If the superintendent of the building knows how to

COMBINATION STACK LINES, AKA COMBO STACK LINE

shut the water that services the line of apartments having the backup, have them do so immediately.

Now, when you get to the job, you will be removing the toilet that is backing up, and cleaning the toilet stack line. If you are on the lowest level of a home or building, then you will be removing the toilet and cleaning the toilet branch line (aka sewer branch line). Also, if you are on the lowest level, it's always wise to check the main house trap before removing a toilet, as you may have a sewer line stoppage.

If you have to, remove the toilet and clear the stoppage from a toilet branch line. Once the stoppage is cleared, reset the toilet and flush a few times to make sure the toilet is working properly.

NOTE: A toilet stack line stoppage may cause a bathtub waste line stoppage to occur, as debris that was backing up in the bathtub may have gotten lodged in the bathtub trap, preventing proper drainage.

Drain Cleaning Cable Sizes and Where They Are Used:

Since you will be using two different drain cleaning drum machines—the Ridgid® K-3800 and K-7500—I will explain which cable sizes and attachments are best for clearing the many different types of drain line stoppages.

Ridgid® K-3800: This drum machine uses four different-sized cables, which can be used in unclogging many different drain line stoppages, which are:

¼" or 5/16" cable – Both of these cables are used in the sink drum. I use the ¼" cable for clearing bathtubs, bathroom sinks, shower stalls, and even kitchen sink line stoppages. (The ¼" x 40' hollow core cable is my preferred choice.) I also cut off the spring tip on the cable and give it a slight hook bend so that it is easier to access a stack line or pass through a bathtub trap.

3/8" cable – This cable is used in the standard-size drum. I use it to clear kitchen sink lines, yard drains, and floor drains (all lines that are two to three inches in diameter). My preferred choice is a 3/8" x 50' hollow core cable with a threaded male

DRAIN CLEANING CABLE SIZES AND WHERE THEY ARE USED

end and a small chisel tip. This cable can also be used with a drop head, small grease spring or any small-sized blade. **Do not use this cable in bathtubs or shower stall waste lines.**

½" cable – This cable is also used in a standard-size drum, and I use it for clearing yard drains, floor drains, main sewer line trap stoppages, and toilet stack lines (any line from three to four inches in diameter). My preferred choice is a ½" x 75' hollow core cable with a male end / cxa adapter and a 1½" single blade, which also can be used with a large chisel head, drop head, or different-sized blades not exceeding three inches.

9/16" cable – I use this cable in the **K-7500**. It is designed for clearing stoppages in lines that are four to eight inches in diameter, mainly for cleaning sewer lines, sewer stack lines, and toilet branch lines. My preferred choice is a 9/16" x 100' hollow core cable with male ends, and my choice of blade to start is a 3" single half blade to break through the stoppage; then I usually change to a 3" (or greater) double blade, depending on the size of the line that I am working on.

For all drain cleaning professionals, it is wise to carry a second drum for the K-7500 machine with an extra 9/16" x 100' cable or even a cable of a thicker size, say, 5/8" x 100', just in case you have a line that is over 100 feet in length. You should start with the smaller cable (the 9/16"), and when you get to the end, disconnect the cable from the drum and remove the drum from the machine. Then hook up the second drum with the 5/8" or larger-sized cable and connect to the 9/16" and continue to clean the line. Running a larger cable behind a slightly smaller cable will help push the smaller cable through the line, providing a better cleaning.

Recommended Hand Tools:

This is a list of the basic hand tools and drain cleaning accessories, that you will need when attempting to diagnose and unclog different types of drain line stoppages.

- 1 – pair of 12" straight jaw pliers
- 1 – pair of 9.5" straight jaw pliers
- 1 – pair of 6.5" straight jaw pliers
- 1 – 6-in-1 screwdriver
- 1 – flat head screwdriver
- 1 – phillips head screwdriver
- 1 – 14" pipe wrench
- 1 – 3-pound lump hammer
- 1 – 1" x 12" flat cold chisel with hand guard
- 1 – 3/4" x 12" bull point chisel with hand guard
- 1 – toilet bowl plunger, aka "the plumber's helper"
- 1 – needle-nose pliers
- 1 – small tubing cutter to cut up to 1½" tubular piping
- 1 – flashlight
- 1 – small 3- to 5-gallon Shop-Vac
- 1 – 12' retractable tape measure
- 1 – pair of ugly gloves
- 1 – pair of safety glasses or goggles

RECOMMENDED HAND TOOLS

1 – 3/8" variable speed corded hand drill with a keyless chuck, #4 - 5/16" combined drill + countersink high speed steel bit, ¼" round arbor with 7/8" bimetal hole saw

Boiler plugs

1 – General T6FL Six-foot Teletube® bulb head toilet auger

1 – set of canvas flush bags 1 1/2" thru 4"

This concludes everything that you will need to know on how-to diagnose, unclog, and help prevent common drain line stoppages.

For more information regarding

- Drain cleaning equipment, tools, and accessories
- Our complete line of eco-friendly sewer and drain maintenance products
- DIY drain cleaning packages
- 1-on-1 hands-on training program

Send us an e-mail by visiting us online at
www.marinisewceranddrain.com or www.marinicorp.com
We also offer **<u>24-Hour Live Telephone Support</u>**

- Perhaps you are performing a drain cleaning task, run into a complication, and are in need of a quick professional opinion or some advice.
- Maybe you have additional questions that weren't answered for you in this book.
- Or you may just want to pick my brain for any additional information.

If so, I am available to take your call 24 hours a day / 7 days a week at: 1-888-INGENIO / 1-888-464-3646 Ext: 04891228

Please be advised that a per minute charge will be billed to your credit card.

Live Chat also available online at:
http://www.ingenio.com/DIYDrainCleaning

Thank you

Tony Marini – Drainage Specialist

CPSIA information can be obtained
at www.ICGtesting.com
Printed in the USA
BVOW09s0748300418
514820BV00015B/368/P